CONTENTS

PAPER TIGERS

The German Armor Crisis 1943 – 1944

by

Aria Lapine

Copyright 2018

'Maybe you should write a book about tanks?'
~ Brian Carman

"The engine of the panzer is its weapon as much as its cannon."

~ Hans Guderian

INTRODUCTION

If a person were to look at the number of words written about armored fighting vehicles, it's fair to say the largest share would be dedicated to describing the operations of Panther and Tiger tanks during World War II. In tactical wargames they appear often, and when they do they are the centerpiece of the German order of battle. I once joked with a friend of mine who plays Advanced Squad Leader that there were probably as many Tiger II heavy tanks included with the complete set as were produced during the war. In the many after action accounts published that feature these vehicles, they are highly effective and universally appreciated by the soldiers that manned them. As much as they were loved by their crews, they were feared by the allies in both east and west. Despite all this praise, the vehicles themselves were in the end a detriment to the German war effort for the simple reason that not enough of them, or none at all, could get to where they were needed.

On the other hand, if there was an award given for the least appreciated armored vehicle of World War II, it would probably go to the SU-76. This 12 ton assault gun was based on the chassis of the T-70, an admittedly poor design from the desperate days of 1942 when Soviet industry followed a 'it's better than nothing' philosophy to their light tank designs before doing the sensible thing and abandoning the light tank concept for any use except reconnaissance. When the more desirable T-34 became available in sufficient quantities to equip Soviet tank brigades, the T-70's design was altered into the much more usable SU-76 by lengthening the chassis, adding a road wheel and replacing the turreted 45mm gun with an armored casement containing the same 76mm gun found on the T-34.

The SU-76 -unloved, but ultimately effective. (RGAKFD)

This is not a vehicle that is represented often in wargames – not many even know of its existence.

Under-armored since it was built on a light tank chassis, it was not liked by its crews, who called it 'The Bitch' and it did not inspire any special fear or awe from its enemies. Because of this, not much has been written about the SU-76 or the battles it participated in, even though over 14,000 were produced and it undoubtedly had a greater impact on the outcome of World War II than either the Tiger or Panther. The reason for this is that it could get where it needed to and provide support to the units it was attached to in sufficient quantities to make a difference. Mere utility, however, does not excite the imagination. While it's true that a Panther or Tiger could destroy an SU-76 in most encounters, what's missing from such an equation is the likelihood of such an encounter ever taking place. Soviet vehicles could go through their entire service careers without ever encountering one of the German heavies.

How did such a mediocre vehicle exert such a profound influence? A large part of the answer is the sheer number produced. The quote from Stalin, 'Quantity has a quality all its own' is often used in a dismissive sort of way, as if to imply the allies in general and the Soviet Union in particular only won World War II through massive quantities of materiel that was of questionable worth. They are correct in the sense that mass conquers all, but how does a nation achieve such an unstoppable force? In short, mere numbers do not equal mass. Mass is achieved by an army with a sufficient quantity of adequate, reliable vehicles that can be sustained through long operations by good logistics and communications. It must have superior leadership capable of wielding it and political leadership capable of giving it realistic goals. The vehicles must also be supported by artillery, infantry, engineers and mechanics that can swiftly return damaged vehicles back to combat. Simply having massive quantities of vehicles accomplishes nothing, as demonstrated by the Soviet Union during the disastrous Barbarossa campaign of 1941. Creation and proper application of mass is an art – achieving it means victory, failure means defeat. The ideal tank, then, will be one that can fulfill the requirements of mass – it first and foremost must be reliable, easy to repair and recover, be able to traverse difficult terrain and have long range and be fuel efficient. Once these qualities are met, the best possible armor and gun can be fitted, but only to the point where the tank's decisive qualities aren't compromised.

This short book could have been longer. Missing are biographies of the military, political and industry leaders mentioned in the text. The reader will also not find much in the way of maps or lengthy descriptions or first person accounts of the many battles mentioned. Such descriptions as they are tend to focus only on the role of Tiger and Panther tanks in these engagements, with any further detail seeming to be digression since this account is primarily concerned with communicating with readers who are familiar with the history of the Second World War. Rather than spill more ink on topics that readers are already acquainted with, I've decided to leave them out. The first chapters of this book will deal with the creation of the first German heavy tank to be mass produced, the Tiger Mk. VI. Through examples, I will try to demonstrate that the Germans should have known that they had a liability on their hands due to the vehicle's overall poor performance. After this prelude, I will move on to the battle of Kursk, where the Germans not only manufactured more Tigers to take part in the offensive, but introduced a newer, even less reliable design in the form of the Panther Mk. V. The poor showing of these vehicles in this offensive and the subsequent Red Army counter-attack should have sealed their fate, but the Germans once again doubled down on heavy tank designs, leading to the loss of much of the Ukraine.

These vehicles did no better against the Western Allied powers, failing to dislodge their bridgehead at Anzio and proving unable to drive them from Normandy. Once the second front had been opened against them, Germany was doomed. Given this fact and that the problems with Germany heavy tanks didn't change significantly for the remaining year of the war, my study of them ends with the Battle of Normandy.

To be fair, I also cover defensive successes in which Tigers and Panthers played a major role, including the German victories in the First Battle of Jassy and the defense of the Panther Line.

The rest of the book undertakes a quick study of Allied armor with an emphasis on the qualities that made them war winners.

CHAPTER 1: THE BEGINNINGS OF HEAVY TANK PRODUCTION

Anyone who has studied tanks is familiar with the idea that the three factors that influence the design of these vehicles are firepower, protection and mobility. Not explicitly stated in the quality of mobility is the concept of reliability. A tank cannot influence the outcome of a battle if it can't get where it needs to go because of a mechanical breakdown, is impeded by terrain it can't cross or runs out of fuel before it arrives. In short, the best armor and gun in the world are useless if the vehicle of which they are a part can't get to where it needs to be.

It has been popular in entertainment media to portray mid to late war German armor – namely the Tiger Mk VI and the Panther Mk V – as nearly invincible weapons manned by elite crews who would confront and defeat massive numbers of inferior allied armor. In operational board and computer games, entire battalions of Tigers and Panthers race across the hexagonal map board, covering dozens if not hundreds of kilometers, without suffering any degradation of combat capability due to vehicles breaking down, something that would have made real life panzer commanders weep with joy. Even in respectable historical publications, one comes across such stock phrases as 'inferior allied armor' or 'the T-34s/Shermans were no match for the heavier German vehicles.' Yet mysteriously, the allied vehicles prevailed – more often than not on the tactical level, and decisively so on the operational and strategic levels of warfare.

How did this happen? The Germans did it partially on their own, on the basis of their flawed designs and their refusal to focus on logistics and the rapidly changing field of combat engineering. They chose to create new vehicles at the expense of proven designs – new vehicles that were complex, difficult to manufacture, prone to breakdown, a challenge to repair and possessing low gas mileage.

The Tiger was the first German attempt to find a technical solution to a perceived armor crisis. The idea that this crisis existed came about due to occasional tactical defeats of lighter German tanks by heavier allied armor, namely the Char B1 bis, Matilda II, T-34 and KV tanks. That these small defeats had little effect on the outcome of the campaigns they occurred in seems to have gone unnoticed.

Still, German technical competency had been challenged and this needed to be addressed, the first example of this new direction in German armor design being the Pzkw VI Tiger. This new design had the deck stacked against it from the beginning, as its designer was none other than Ferdinand Porsche. As a member of the SS and Hitler's inner circle, Porsche was a Nazi party luminary. Hitler also considered him a great German engineer, which made it easy for Porsche to convince him to put him in charge of tank production, but failing at this task of industrial organization, tried his hand at designing tanks instead.

Unfortunately for the Germans, Porsche turned out to be as bad at tank design as he was at organizing production. Even though the design was rejected in favor of a more reliable one put forward by the firm of Henschel, he still managed to convince Hitler to produce 90 chassis of his design. These hulls would eventually be the basis for the Ferdinand tank destroyer, but the production of these and other Tiger prototypes disrupted German armor production for months during the crucial year of 1942. The Nibelungenwerke facility in Austria where they were produced had originally been intended to manufacture the Pzkw IV – due the distraction, only 186 examples of this vehicle were manufactured in 1942 at this factory instead of the 1800 that could have been produced. In exchange for a proven design that could been used to replace obsolete vehicles in German panzer divisions, the Heer instead got a barely mobile, unreliable vehicle that would generally break down only after a day or two of combat operations.

When the Tiger was first used in combat around Leningrad in 1942, the Germans should have realized that it was a poor substitute for the tanks that could have been manufactured in its stead. Throughout the siege, a maximum 33% of the Tigers of Heavy Battalion 502 were ready for operations at any given time, with sometimes as few as two available. During operations, Tigers would often get bogged down in the marshy terrain that surrounded the city – due to their weight and the lack of suitable recovery vehicles, once they were stuck, they were nearly impossible to extract.

Tiger I abandoned near Leningrad. (Bundesarchiv)

Operation Spark, the Soviet attack that finally opened a corridor to the besieged city, offers an example of a common Tiger engagement during the fighting that consumed the city and its environs until the siege was finally ended in 1944. Launched in January 1943, Operation Spark was an offensive that finally achieved a measure of success after several previous failures. Operating as the main operational reserve for the 26th Corps, Heavy Battalion 502 had 7 Tigers and 16 Pzkw IIIN available to support the defense. While undertaking a counter-attack, Battalion 502 not only failed to stop the Soviets from regaining contact with the city, but became trapped in the developing offensive. All of the Tigers were lost, including one that drove into a peat bog and was captured intact after its crew fled when Soviet infantry took them under fire. The Germans tried in vain to recapture the vehicle, but only succeeded in losing several Pzkw IIIN during the attempt.

The Pzkw IIIN – this vehicle was used to fill out the order of battle of the Tiger battalions until more of the heavy tanks became available. It was used for fire support, reconnaissance and even as a liaison vehicle. They were phased out in 1943 even though many heavy battalion commanders found them useful.(Bundesarchiv)

Ironically, the much maligned Soviet T-60 light tank did far better than the Tiger in this particular operation, its light weight allowing it to cross terrain the Tiger couldn't, lending valuable support to the infantry units it accompanied. To make a contribution to any battle, a tank has to first and foremost be able to show up in sufficient numbers to make a difference.

The Soviet T-60 – a tank in hand is worth more than any number stuck in the mud kilometers away from the battlefield (RGAKFD)

While Tigers were instrumental in halting some Soviet attacks around Leningrad, a battalion of Stug IIIs armed with the powerful 75mm L/46 would have been a more effective solution – this weapon outranged and could destroy all Soviet tanks from any aspect until the JS series of heavy tanks appeared in 1944. It's better reliability (80% on average) would have meant larger numbers of effective weapons engaging Soviet armor, leading to better overall outcomes to German operations.

The Stug III, which was based on the chassis of the Pzkw III. While manufactured in large numbers, there were never enough of them.(Bundesarchiv)

The Stug III's lighter weight was also better suited to the soft ground of the region and, unlike the Tiger, could also use most bridges. If a Stug did break down, it could be easily retrieved by the German's standard recovery vehicle, and once brought back to the depot, could be quickly repaired due to its simpler design – it even shared many components with the Pzkw III. In the above example, a Stug battalion with infantry and artillery support probably would have succeeded where the Tigers failed.

Meanwhile, Tigers were also being used far to the south, where the Germans were trying to prevent the Soviets from taking Rostov before the 4th Panzer Army could be extracted from the Caucus region. On January 9th, 1943, the 23rd Panzer Division, supported by 17 Tigers from Heavy Battalion 503 counter-attacked the Soviet 3rd Mechanized Corps. Despite going in four times, the Germans failed to dislodge or even seriously harm the defenders. When it was over two days later, Heavy Battalion 503 had lost fifteen tanks to defensive fire and breakdowns, while the 23rd Panzer Division was down to 7 operational tanks.

Meanwhile, Heavy Battalion 503's 3rd company was supporting the 19th Panzer Division during its withdrawal. It had destroyed 11 Soviet tanks during its engagements, but was down to its last 3 Tigers. Both the 19th and 23rd Panzer pulled back behind the Manych River on January 16th, uniting the separated elements of Heavy Battalion 503. After only a week of combat, it was down to only two Tigers and had to be rebuilt back in Germany.

CHAPTER 2: NORTH AFRICA

After its combat debut in the Soviet Union, the Tiger was introduced into a very different theater of war – North Africa. Heavy Battalion 501 was sent to Tunisia along with other units to bolster the western flank of the Afrika Corps after Western Allied landings in Morocco and Algeria; it was able to make only a minimal contribution to the defense due to low numbers, poor tactical handling and the usual problem of vehicle reliability. The battalion had 22 Tigers assigned to it in its order of battle, but the most it had operational at any one time was 14.

Tiger from Heavy Battalion 501 in N Africa. (Bundesarchiv)

The unit's first encounters were with lighter Crusader and Stuart tanks (as well as a few M3 mediums) – while these engagements invariably resulted in a victory for the Tiger, similar and perhaps more decisive results could have been had with greater numbers of Pzkw IVGs or even Pzkw IIIJs with their long 50mm guns.

The Tiger's first major combat action was Operation Eilbote I in January 1943. Meant to clear the lines of communication from Tunis and Bizerte to Rommel's HQ, it was undertaken by the now fully assembled Heavy Battalion 501, whose 2nd Company had finally arrived in Tunisia. Tasked with supporting Battle Groups Weber and Luder, in this particular battle the Tiger proved effective, with its heavy armor being proof against AT guns that might have halted lighter vehicles. All the same, four Tigers were lost, with more damaged by mines, direct and indirect fire. By the time the operation ended, only one Tiger was still operational, with only three Tigers on average being available on any given day. Despite these losses, the Tigers had achieved an operational objective that might not have possible without their assistance.

The Pzkw IIIJ – while not as heavy as the Tiger, its 50mm gun was potent enough to challenge any Allied tank in North Africa. Greater numbers of this vehicle would have had a more positive impact on Axis operations than the presence of a few Tiger tanks. (Author's collection)

The next North African engagement the Tiger participated in was Operation Fruhlingswind in support of the 5th Panzer Army's attack on Sidi Bou Zid. Only the 1st Company (with six Tigers available) took part in the battle, but was nonetheless an essential element in its success. Leading Battle Group Reiman, the Tigers proved to be impervious to American defensive fire. After attacking and surrounding the American strongpoints, the Tigers assisted in driving off two companies of the American 1st Armored Regiment along with a company of M-10 Tank Destroyers which had arrived to relieve the defenders. Later on, the Tigers helped defeat a counter-attack by M4 Sherman tanks from the same regiment.

The final major North African engagement involving Tigers was Operation Ochsenkopf. This offensive was meant to extend the bridgehead of the 5th Panzer Army in Tunisia and targeted Beja. The heavy tanks were part of Kampfgruppe Lang that included 77 tanks, 14 of which were Tigers.

Even early in its career, the Tiger wasn't immune to shell fire. (Bundesarchiv)

The first opposition encountered was the 5th Battalion of 128th Hampshire Infantry Brigade, supported by numerous batteries of 25 Pounder artillery guns of the 172nd Field Regiment and 155th Battery at Sidi Nsir. This small force was able to delay the Germans with minefields and direct fire from the 25 Pounders.

While the British outpost was eventually taken, it had delayed the Germans for twelve valuable hours. This time was put to good use by the main British position at Hunt's Gap 20 kilometers to the south-west. The remainder of the 128th Brigade was dug in here, supported by 72 25 Pounders and 15 5.5 inch guns. Two squadrons of Churchill Mk IIIs had arrived as reinforcements and a squadron of Hurricane Mk IVD fighters equipped with under wing 40mm cannons was also in the area.

Muddy ground forced the Tigers and other vehicles to stay on the road or face bogging in the soft ground on either side. The British, meanwhile, had placed their Churchill tanks hull down, supplementing these with several AT guns as well as medium and heavy artillery firing over open sights.

The Churchill Mk III in N. Africa (National Archives, UK)

The Germans attacked through the night of the 27th of February, but by the next day seven Tigers sat immobilized in the minefield with the British defense still intact. Of the fourteen Tigers that started the operation, only two were still in running order with several destroyed by British sappers as they sat stranded. With the British artillery keeping the German infantry at bay, none of them could be recovered. On March 2nd, the Germans withdrew having lost over 40 tanks total as well as 60 other armored vehicles.

In both the Soviet Union and North Africa the Tiger could be useful at the local, tactical level, but less so operationally. It could win engagements, but not battles because of the low numbers available due to their complexity which made them both difficult to manufacture and maintain in the field. The German's logistical difficulties exacerbated these problems by making it challenging to get spare parts where they were needed, which often lead to tanks being cannibalized in order to repair others. Strategically, they made no difference at all – the campaigns they participated in were not decided due to their presence or absence.

The logical choice going into 1943 would have been to cease manufacturing the vehicle altogether, just as the Soviets had done with their impressive but unworkable KV-2. As we shall see, the Germans did not follow this prudent course, producing the Tiger throughout 1943 at the expense of more reliable and effective designs. This decision would cost them dearly at the battles of Kursk and the Ukraine that would be fought later in that decisive year.

The massive KV-2 – in 1941 the Soviets realized this tank wasn't suitable for the sort of mobile war they would be fighting and stopped manufacturing it. (Bundesarchiv)

CHAPTER 3: GERMAN HEAVY TANKS AT KURSK

At the start of the Battle of Kursk there were 147 Tigers assigned to various formations participating in the battle. Two army level formations, Heavy Battalion 505 in the north and 503 in the south had an assigned strength of 45 tanks each. These were the later Organization E units which did away the lighter Pzkw IIIN tanks in order to have a formation composed entirely of the heavier Tiger. Each of the three SS Panzer Divisions assigned to Army Group South's 4th Panzer Army had a company of Tigers assigned with a strength of 14 tanks each, as did the mechanized Gross Deutchland Division. Mechanical attrition and other problems such as lack of adequate bridging reduced the total number of Tigers available to 97 before the first shots had even been fired. The numbers of operational vehicles fell even more rapidly after the fighting began, with many falling prey to mines that could not be cleared by engineers due to heavy artillery fire. By the time the offensive ended, the two battalions had only eight tanks listed as total write-offs. However, 58 were damaged or suffered breakdowns, leaving only 24 fully operational. Some of the disabled tanks were later repaired, but many were eventually written off.

While resistant to shot from most angles in 1943, the Tiger was more vulnerable to less conventional forms of attack. Here a Tiger undergoes repair at Kursk after suffering mine damage. (Bundesarchiv)

In exchange, the two battalions claimed 182 Soviet tanks destroyed, but like most claims made in the heat of battle were overstated. German intelligence would often reduce such claims by half in order to arrive at a more realistic number of enemy losses. In addition, many of the Soviet tanks that were hit were only damaged and were later recovered, and these had a much better chance of being sent back into action quickly than the mechanically temperamental Tiger. The vast majority of German tank kills at Kursk were done by the workhorse of the Panzer arm, the Pzkw IV.

The Pzkw IV at Kursk. This vehicle and its stablemates, the Pzkw III and Stug III made up the vast majority of German armor at Kursk and contributed most to what success it was able to achieve.(Bundesarchiv)

The experience of Heavy Battalion 505 in the 9th Army on the northern shoulder of the Kursk bulge highlights another problem when reading about the effectiveness of German heavy tanks – namely, crediting them solely with a tactical victory when in fact they had only lent support to the combined arms team that had won the battle. For Heavy Battalion 505, this excessive praise came in two parts – one, crediting it with an operational victory when only a limited tactical success had been achieved, and secondly not mentioning the other combat elements present that had actually made that victory possible.

Some accounts of the opening engagements of the German 6th Infantry Division at Kursk credit this Tiger battalion with single handedly fighting its way through the Soviet's first defensive belt, manned by a rifle regiment of the 15th Infantry Division. These accounts go on to claim that a breakthrough to Kursk itself was only prevented by the absence of panzer divisions needed to exploit the breakthrough.

What follows is a more detailed view of what actually happened. The Germans used B IV demolition vehicles to try and open a breach through the Soviet minefield that screened the Soviet advanced positions. These experienced difficulty in completing their task – a complete breach failed to materialize and six Tigers were immobilized in the minefield, leaving 25 to continue the fight.

German B IV demolition vehicles. These were designed to breach minefields and destroy other fortifications with their high explosive payloads but were too thinly armored to have much chance of success. The Germans would have been better served attaching a roller or flail onto one of their battle tanks like the allies did. Note the two Ferdinands in the background. (Bundesarchiv)

The Soviet rifle regiment, meanwhile, was in a state of confusion due the German artillery preparation and had lost contact with its divisional HQ. Exploiting this opportunity, the 6th Infantry Division moved forward and created a breach in its defenses. After this was accomplished, the 20th Panzer Division was sent forward to exploit the gap, taking out one of the regiment's battalion strongpoints in the process. While the 15th Rifle Division's first line of defense had been compromised, the second was still intact and the Soviets sent T-34s of the 237th Tank Regiment as well as the 1441st SP Artillery Regiment to block any further advance. While Walter Model, commander of the 9th Army, sent the 2nd and 9th Panzer Divisions to assist in reducing the Soviet defenses, these would not arrive until the following day.

Even so, behind the 15th Rifle Division was the intact 6th Guards Rifle Division, fresh and still firmly entrenched in their positions. In addition, General Rokossovsky of the Soviet Central Front still had many more untapped reserves to commit. No breakthrough had been achieved, although a Soviet rifle regiment had been mauled in the fighting. Even this tactical success had only been achieved with a combined arms attack that involved the efforts of several battalions of artillery, an infantry division, 2 Stug III assault gun battalions, a pioneer battalion, a panzer division . . .and twenty-four Tiger tanks that had actually contributed little to the battle.

Meanwhile, the southern shoulder of the Kursk salient was seeing the combat debut of Germany's new "medium" tank, the Pzkw V Panther. The Panther was a medium tank in name only, weighing in at 44 tons. The T-34, its chief opponent at Kursk, came in at a mere 30 tons. It's development history was clouded by corruption, the result of which was a tank that incorporated several design features wanted by Nazi Party bureaucratic 'experts' instead of those requested by soldiers doing the actual fighting.

The design was first proposed due to German encounters with Soviet T-34 and KV tanks in 1941. In November of that year, an armor investigation committee visited General Guderian, a luminary of Blitzkrieg tactics who was currently leading the 2nd Panzer Army in its attack on Moscow. As a result of this meeting, a 30 ton tank design was proposed with 60mm of frontal armor and a gun capable of defeating the latest Soviet designs.

A classic example of design overreach, the Panther was given features that were technologically advanced but mechanically unsound. (Author's collection)

Two German manufacturing companies, Daimler-Benz (D-B) and MAN, began developing competing designs to meet these requirements. D-B submitted a design that used many of the notable mobility features of the T-34, including a rear wheel drive and a diesel engine. MAN, meanwhile, went with a gasoline engine that was already in production, wedded to a less efficient front-wheel drive.

Though MAN's design was inferior, it had an ally in the form of one Heinrich Kniepkamp, a former employee who now was in charge of all tank projects in the German Army Ordinance Department. Despite Kniepkamp's influence, Hitler initially favored the Daimler-Benz proposal, finding merit in its fuel efficient diesel engine.

To head off this setback to his personal ambitions, Kniepkamp committed industrial espionage by sending details of the D-B design to MAN. He even went so far as trying to influence Nazi officials by saying it looked too much like the T-34, an inferior Slavic design that might make the world question German technical competence. This blatant pandering still wasn't enough to sway Hitler's judgment in favor of MAN's design.

It took the interference of another ambitious Nazi engineer, Karl Saur, to convince both Armaments Minister Albert Speer and Hitler that developing D-B's diesel engine would take too long, and that a new tank design had to be put into the hands of the German military as soon as possible.

As if this weren't enough, Hitler himself made the decision that ended up making the tank a battlefield liability by insisting that the frontal armor be increased from sixty to eighty millimeters, which increased its weight by 15 tons. To be fair, it wasn't entirely Hitler's fault – he wanted the design to incorporate a certain useful feature, and it was up to the engineers to tell him that incorporating it compromised the vehicle's design to a dangerous degree. Fearing the loss of a valuable contract, they remained silent.

Even this support was nearly overturned by the new vehicle's disastrous performance during a demonstration in January 1943 for Albert Speer. Six of the thirteen vehicles broke down, one of them even catching on fire. Instead of canceling the project outright, the vehicles were sent back to the factory for further rebuilding. In June, mere weeks before the vehicle's debut at Kursk, further testing revealed that the vehicles still weren't ready and the entire first production run of 250 vehicles was sent back to the factory yet again for rebuilding.

Hans Guderian, in his role as Inspector of Armored Vehicles, tried to put a stop to the Panther's deployment, but was shut down by Saur who knew that cancellation of the program at this point would be disastrous for his career. To ensure the program's continuation, Saur made sure the Panther never underwent serious field testing.

While the Panther did have a gun that excelled at knocking out armored vehicles (which was perhaps more suited to a specialist tank destroyer as it had mediocre HE performance) and armor on the frontal aspect that was proof against the most common Allied AT weapons, in the end its production drained scarce resources from production of the Pzkw IV and Stug III.

As it turned out, the Panther's tortuous development was an omen for its disastrous introduction to combat. Two hundred Panthers were available at the start of the Kursk offensive, assigned to the 39th Regiment of the Gross Deutschland (GD) Division. After being off-loaded from the rail cars that had delivered them, two Panthers caught fire and burned themselves out and eighteen others broke down during the short drive to their assembly area. During their run in to their attack sector, four more suffered from leaky fuel lines, burst into flames and had to be abandoned.

Panthers being delivered to Kursk. Even after multiple rebuilds, the vehicle still wasn't ready for combat. (Bundesarchiv)

The plan was for the Panthers to break through the first line of Soviet defenses, which would allow the GD Division to advance to its objectives. However, when the Panthers approached the forward edge of the battlefield, they found a minefield and wire obstacle in front of them and an eighty meter wide ravine that had gone unnoticed due to poor reconnaissance. Having had no time to train because of their late arriving vehicles, the Panthers made a grave error – instead of maintaining a safe interval between them, they crowded forward in front of the obstacle, bunching together and making an excellent target.

As it turned out, the GD's combat engineers had been at the ravine but had determined that it couldn't be crossed due to the thick mud that covered the bottom. Having gone off to find an alternate route, they had left no word of their conclusion or where they had gone.

Impatient to move forward, several Panthers tried to cross through narrow lanes cleared through the mines, but all of them ended up bogged down in the bottom of the ravine. Another company tried to find a different crossing point but drove into an undiscovered minefield. In little time, twenty-five Panthers had been immobilized due to mines, mud and breakdowns due to their temperamental engines and weak final drives. Just when it seemed things couldn't get worse, artillery shells began falling among the bunched up, immobilized armor, killing six crewmen, destroying another tank and damaging several more. By the end of the day only eighty Panthers, possibly far less, were still operational. Recovery efforts for Panthers bogged down in the ravine were underway, but were hindered by the vehicle's enormous weight.

*A Panther with a tow bar attached, abandoned on the Kursk battlefield. (RGAKFD)*

The disaster continued the next day when the remaining Panthers in the regiment got lost and drove into an ambush. Many were immobilized by mines, while others came under fire from dozens of dug in T-34s. In the process of withdrawing, nineteen more Panthers were lost. Eventually, the unit was able to locate the rest of the GD Division's armor but didn't manage to accomplish anything else.

On the morning of day three of the offensive, only 50 Panthers remained.

Soviet soldiers inspect a Panther destroyed during the battle of Kursk. While the vehicle had superior armor protection from the front, it was still very vulnerable from the sides and rear. (RGAKFD)

The Tigers also failed to have much of an impact in the south. Heavy Battalion 503 was attached to the 3rd Panzer Corps, who gave one of the battalion's companies to each of its panzer divisions. The heavy tanks had to spend most of the first day of the offensive waiting for engineers to construct a bridge that could bear their weight. After they crossed, the 2nd Company lost thirteen of its fourteen tanks in a minefield. While these tanks could be repaired, it would take time and until that occurred, they were out of the fight.

While much is made of the low number of Tigers destroyed during Kursk, what isn't mentioned is that very few of them managed to make it to the thick of the battle at all. The 3rd Panzer Corps was attached to Army Detachment Kempf, whose mission was to protect the flank of the 2nd SS Panzer Corps as it battled its way north. By July 7th, the second day of the offensive, Kempf was well behind the northern most spearheads of the SS, exposing these units to numerous counterattacks along their eastern flank. During the battle, Heavy Battalion 503 was attached to Panzer Regiment 11 of the 6th Panzer Division, whose mission was to infiltrate Soviet lines and capture bridges over the Donets River. Kept at the rear of the column to ensure a breakdown didn't block the road, the Tigers contributed little. When the Soviets detected the Germans' presence, they blew up the bridges and the column had to wait days for engineers to bring another one up that would be capable of supporting the Tigers' weight. While they were eventually able to seize a bridge over the North Donets River and link up with the SS spearhead, by this time the climax of Kursk, the Battle of Pokharovka, had been fought and the German offensive had been called off.

Despite the failure of the Tiger and Panther tanks to significantly affect the outcome of Kursk, narratives of the day to day action often focus on the few tactical victories these vehicles were present for, downplaying their operational insignificance. They not only failed to justify the huge amounts of resources spent on their creation and production, they ended up being a burden to the units they were supposed to be supporting due to their inability to use most bridges, the difficulty in recovering and repairing them once damaged and their small numbers on the battlefield which meant that the units relying on them for support were often left without it.

While the weather would have prevented a German assault in May (the battlefield was a sea of mud at this time), an attack in June would have been possible if not for the delay in waiting for 'sufficient numbers' of Tigers and Panthers to be available. The Soviets, meanwhile, used their time to good effect by fortifying the Kursk salient and building up the mass of reliable vehicles that would ultimately win the day.

The Tiger continued to prove itself a useful defensive vehicle in some cases as long as the front was fairly stable, as proved by Heavy Battalion 503 when it assisted the 4th Panzer and 6th Infantry Divisions in holding off attacks by the 2nd Tank Army during its attack toward Orel. Like Leningrad, defending a static position where its limited mobility and lack of reliability weren't as much of an issue because of the short distances involved, brought the Tiger's strengths of strong armor and excellent gun to the forefront. However, even with eight panzer divisions (half of those available on the Eastern Front), the Germans could not hold Orel, the Soviets taking the city only one month after the Germans had started their Kursk offensive.

While a tracked mount for the formidable 88mm gun was desirable, the Germans would have been better off going with a lighter weight alternative than the Tiger. (Author's collection)

CHAPTER 4: ACROSS THE DNEPR TO THE UKRAINE

Meanwhile, Army Group South would also face the test of a Soviet offensive, Operation Rumyantsev, which took place between August 3rd – 23rd. After the failure of their attack at Kursk, the 4th Panzer Army had withdrawn to its start lines and had since had much of its remaining tanks and other combat elements sent to other sectors of the front that were weathering Soviet offensives, the 2nd SS Panzer Corps being sent south to the Mius River battles and the Gross Deutschland Division sent north to assist the 9th Army in its defense of Orel. The only armor initially available to defend Belgorod and Kharkov were the 19th and 6th Panzer Divisions with less than 100 operational tanks between them. Also available were 27 Panthers belonging to the 52nd Tank Battalion (with another 106 awaiting maintenance or repair). Heavy Battalion 503 was down to a mere six tanks, with another 32 in the repair depots.

Meanwhile, many of the Soviet tanks that the Germans had counted as destroyed had been repaired and were at that moment preparing an attempt to drive their enemy across the Dnepr River. At 5 am August 3rd, combat engineers of the Voronezh Front advanced with the combined support of three artillery divisions, whose fire practically destroyed the 52nd Corps' defense line. After only three hours of combat, General Vatutin was able to commit his exploitation force of two tank armies through a breach in the German lines.

The Soviets advanced rapidly, bypassing German attempts to create defensive positions, eventually reaching the town of Tomarovka where the Germans were forced to destroy 72 Panthers that were in the repair shops, along with several Tigers.

A Soviet soldier inspects a Panther abandoned during the retreat from Kursk.(RGAKFD)

By August 9th, only five Panthers remained of the 200 that Army Group South had possessed only a month before, along with 96 new Panthers received in mid-July. There were few resources available to stop the continuing Soviet offensive, Vatutin committing his tank armies in echelon to sustain the momentum of his offensive. This method was only possible due to the superior reliability of the T-34, which allowed large numbers of vehicles to remain in their combat formations even after advancing dozens of kilometers. A similar advance would have left a unit equipped with Panthers or Tigers unfit for combat.

The T-34 was very good at the quality that was most important in tank design – the ability to show up in sufficient numbers to make a difference. (RGAKFD)

By early July the 51st Panzer Battalion, re-equipped with Panthers, returned to help defend Kharkov. While this unit, along with Tigers and other German armor, were able to score some tactical successes, lack of sufficient mass due to frequent breakdowns meant that they weren't able to turn these victories into anything operationally meaningful, since at the end of the engagement only a handful of the heavy vehicles would be available for further combat. For example, when the Gross Deutschland Division tried to cut off the spearheads of the 4th and 5th Guards Tank Corps, they were only able to assemble a force of fifteen Panthers and ten Tigers. Attacking jointly with the 3rd SS Panzer Division, they succeeded in isolating the two corps, but had insufficient force to hold the pocket closed, much less eliminate them. By August 24th the two German formations had lost most of their tanks and were driven out of the city of Akhtyrka, an important position that protected the deep northern flank of the forces battling around Kharkov, which itself had fallen on August 23rd.

After the fall of Kharkov, the Germans began a rapid retreat to the Dnepr River. Unable to conduct an effective fighting withdrawal due to a lack of armored vehicles, the Germans lost a great deal of artillery, supplies and 80 more Panthers, most which had suffered breakdowns and needed to be destroyed to keep them from falling into the hands of the enemy. Lack of tanks in the combined arms equation also led to the decimation of German infantry units, dooming the Dnepr defense line before it had even been manned. The Soviets were able to cross the Dnepr almost immediately on reaching it, their initial improvised means soon being supplemented by ferries and pontoon bridges.

A virtual stalemate followed, with the Germans unable to mass sufficient strength to eliminate any of the bridgeheads while the Soviets tried to assemble the needed logistics and force at the correct location in order to take Kiev. Stavka, the Soviet high command, came up with an audacious plan, ordering Vatutin's 1st Ukranian Front to move Rybalko's 3rd Guards Tank Army (GTA) from the bridgehead at Bukrin to Lyuthzh, a march of over 150 kilometers. Rybalko accomplished the task in five days – this was no small feat, as he first had to withdraw his tank army to the east bank of the Dnepr, cross the Desna at Letki, then move it to the west bank again – all at night and with limited bridging available. The night marches, fog and Soviet radio deception kept the Germans from noticing this major adjustment of force.

When the Soviets launched their new offensive, it took the Germans completely by surprise. Caught with only light field works as protection, the divisions of the 13th Infantry Corps were able to hold the line for a day, but in the end almost evaporated under the concentrated fire of the 13th Artillery Corps. The 8th Panzer Division, kept close to the front line as an operational reserve, was also caught unaware and managed only a couple of counter-attacks during the day. Having only fourteen tanks available for operations, it lacked sufficient mass to have any impact. By nightfall of November 4th, the Soviet 51st Rifle Corps was in the suburbs of Kiev.

Meanwhile, the 6th Guards Tank Corps was racing southwest toward the rail station at Fastov, occupying it before the 7th Panzer Division could intervene and disrupting the deployment of the 25th Panzer Division that was due to detrain there. By November 6th, the center of Kiev was occupied by the Soviets, causing a general German withdrawl.

The entire operation to sieze Kiev serves to show the superiority of the T-34 over German armor, the key to this superiority being its ability to maneuver hundreds of kilometers in inclement weather over varied terrain. Its reliability under these conditions allowed the units that employed it to retain sufficient mass to remain viable in an operational context. This operational ability allowed Soviet tank units to operate with such speed that they were able to preempt German counter-measures, usually multiple times before the unit would have to be withdrawn to refit. Having quantity reliably available implies quality in terms of tank design – it is not the only consideration, but it is the most important one.

Not willing to allow the Soviet victory to stand, the Germans began assembling forces for a counter-attack in an attempt to recreate a scenario similar to the defeat of Soviet spearheads around Kharkov in early 1943. Fresh reinforcements were brought in to make this possible, including the 1st SS Panzer, 1st Panzer and 25th Panzer Divisions along with Heavy Battalion 509 with a complement of factory fresh Tiger tanks – these formations were all rested and at full strength. Manstein, who would be leading the attack, also received the 2nd SS Panzer Division which still had 33 tanks available. These armored units, organized under the HQ of the 48th Panzer Corps, had a combined strength of 558 tanks. Manstein's first objective was Fastov, where he would attempt to flank and destroy the 3rd Guards Tank Army.

Tigers in the Ukraine. Despite attempts to improve the reliability of the design, the Tiger was never an operationally viable vehicle. (Bundesarchiv)

Unfortunately for Manstein, the 3rd GTA was not resting on its laurels and was continuing its attacks on the 4th Panzer Army. While he was still assembling his forces, Soviet tanks were approaching the German railheads where his units were offloading, including the vital logistics center of Zhitomir. In an attempt to slow down the Soviet advance, Manstein sent in what forces he had available, the 25th Panzer Division and Heavy Battalion 509 both being fully deployed and available for operations. After several days of attacks, these units not only failed to take Fastov, but the inexperienced 25th Panzer Division broke and fled on its first contact with T-34s, losing many of its vehicles as it did so. Heavy Battalion 509, meanwhile, had only 14 of its 45 Tigers left for the main German counterattack that was still to come.

Just as he was about to begin his offensive, Manstein found himself preempted once again when Vatutin sent the 1st Guards Cavalry Corps and 38th Army to take the vital German rail center and logistics base at Zhitomir, with the city falling on November 13th. Manstein, who was going to focus his offensive at seizing Fastov, instead had to re-orient his forces to retake both cities. Zhitomir did fall to the German attack on November 20th, but once again German forces lacked sufficient armor to seal off their encirclements with more than fifty of the 1st Panzer Division's Panthers suffering breakdowns after only a few days of combat.

Vatutin, meanwhile, had been parrying the German blows with independent tank and anti-tank regiments assigned to his infantry armies, withdrawing the battered 3rd GTA to the reserves to refit.

Panthers move up to the front in the Ukraine. Like the Tiger, improvements were made to the design to try and improve its reliability but they were too little and too late – the Germans needed vehicles capable of reversing their fortunes in 1943 and didn't get them.(Bundesarchiv)

After retaking Zhitomir, the Germans didn't accomplish much, battering their armor against the Soviet screen of infantry corps and anti-tank regiments throughout the month of December. No major Soviet units were destroyed and no important ground was seized, with Kiev itself well behind the Soviet lines throughout the operation. For its efforts, 4th Panzer Army had lost over half of its tanks with only 28 Panthers and 11 Tigers still running. The Soviets, on the other hand, had recovered and repaired many of the damaged tanks belonging to the 3rd GTA, refilling any gaps in its order of battle with replacement vehicles.

In a significant feat of engineering, the Soviets had also managed to repair the rail and road bridges spanning the Dnepr at Kiev in just a few weeks, improving their logistical situation and allowing the 1st GTA to be railed directly across where it joined the quickly refurbished 3rd GTA. On Christmas Eve, these two armies drove the Germans back, passing many broken down Panthers and Tigers that the Germans couldn't recover on the way. By December 30th, Zhitomir had fallen to the Soviets and the German front had once again been shattered.

Bundesarchiv, Bild 101I-664-6759-30
Foto: Schödl (e) | 1943 Herbst

Panthers moving up to the front with a load of infantry. While many of the German counter-attacks in the Ukraine were well conceived, they didn't have the materiel necessary to make them effective or decisive.(Bundesarchiv)

The failure of the German offensive to restore the Dnepr River Line marked Germany's last chance to achieve a stalemate on the Eastern Front. The Panthers and Tigers that were supposed to restore German fortunes instead took resources away from more battle-worthy designs that might have been able to hold that vital river had they been present in greater quantity. While the Panther's high velocity 75mm and the 88mm gun of the Tiger were certainly effective in the anti-armor role, the difficulty in producing these vehicles and maintaining them in the field guaranteed that allied tanks rarely encountered them. The lack of sufficient battle tanks would be felt in German Panzer divisions that had to continue using Pzkw IIIs or Pzkw IVs with the earlier, low velocity 75mm gun. Oftentimes, crews without an available vehicle would be pressed into service as infantry, their valuable technical skills lost when their positions were overrun.

The Soviets continued to move forward despite increasing logistical problems and losses, but were lured forward by the major German supply depot located at Uman. Manstein was able to assemble yet another counter attack force with difficulty by shifting units from General Hube's 1st Panzer Army which was further to the south. This would have repercussions later, as it left General Wohler's 8th Army dangerously unsupported. The counter-attack force consisted of the 1st SS Panzer Division, the 16th Panzer Division, the 101st Jaeger Division and a battle group called Panzer Regiment Bake after its commander. This last unit achieved a certain amount of notoriety during the Eastern Front winter battles, and consisted of 34 Tigers from Heavy Battalion 503 and 46 Panthers and other units from the 6th Panzer Division. Organized under General Vormann's 47th Panzer Corps, the unit was able to halt Vatutin's drive, inflicting considerable losses but not destroying the 1st Guards Tank Army as is so often claimed. In reality, the Soviets were able to perform a fighting withdrawl, holding up the German advance with mines and AT guns. While encirclements did take place, they mostly closed around little or nothing, with many of those trapped able to escape due to the thin cordon around them.

What is surprising is that despite their poor performance, the Germans would continue to manufacture their heavy tanks, even increasing production. This error would go a long way to ensuring the Heer's experience on the Eastern Front in 1944 was an even greater disaster than 1943. The Soviets, meanwhile, would continue to manufacture the T-34, both in an improved version of the original model and the T-34/85 with a new turret and an 85mm gun that was superior to the most common German AT weapons and had a much better HE round than the Panther's high velocity 75mm gun. On a main battle tank, HE performance tended to be more important than AP since infantry, their strongpoints and accompanying AT weapons were far more common targets than tanks. Other designs ended up using the T-34 chassis as well – the SU-122 providing heavy infantry support with its casemate mounted 122mm gun and the SU-85 and SU-100 providing mobile striking power to Soviet AT regiments.

The T-34/85 was a war-winning design, keeping the light weight, reliability and ease of manufacture of the original T-34 while increasing its firepower with a new 85mm gun. This made the vehicle more than a match for its most commonly encountered enemies, the Pzkw IV and Stug III.(RGAKFD)

CHAPTER 5: THE RELIEF OF LENINGRAD AND THE DRIVE TO THE PANTHER LINE

Further to the north, the Soviets opened their year of deadly offensives with a bid to drive the Germans away from the city of Leningrad once and for all. Several other attempts to accomplish this had been costly failures, but this latest effort would begin from the Orienbaum Bridgehead, an isolated pocket of Soviet forces to the west of Leningrad kept supported and supplied by ships from the nearby fortress of Kronstadt during the spring and summer months, or across the ice of the narrow Gulf of Finland when the ice had grown sufficiently thick. The Germans considered this a quiet sector and an unlikely place for an offensive to originate given the logistical difficulties of assembling and supplying a sizable force there. Because of this the German lines were thinly manned, the Soviets being careful to keep their enemies in the dark by only moving forces across by night. Their attempts to conceal their intentions bore fruit, and Germans were caught by surprise - on January 14th, 1944, the 2nd Shock Army was able to rapidly affect a breakthrough. Although the initial advance was rapid, the Germans had anticipated an offensive somewhere along the 18th Army's frontage and had operational reserves available. The 3rd SS Panzer Corps HQ had been sent to Army Group North along with the 11th SS Panzer Grenadier Division and the 4th SS Panzer Grenadier Brigade. The 11th SS had a battalion of Panthers attached but these were immobile due to manufacturing defects. A few of these were emplaced as strongpoints, but all sixty of them had to be abandoned during the subsequent German retreat. The mobile anti-tank defense for the German forces once again relied upon the dependable Stug III, 52 of which were available for operations.

With the Panthers sidelined and few Tigers available, it once again fell to the Stug III to provide mobile AT defense at Leningrad. (Bundesarchiv)

After five days of fighting the Soviets were able to achieve a breakthrough and defeat the German operational reserves, the 2nd Shock Army and the 42nd Army taking the vital road junction of Ropsha by January 19th, which effectively isolated several German formations. With its front lines shattered Army Group North was nonetheless ordered to remain in place by the German high command, who promised reinforcements in the form of Panzer Grenadier Division Feldherrnhalle and the 12th Panzer Division, but the former unit had to make its way from France and the latter had to be detached from Army Group Center. Neither would be available in the near future. The only reserve close at hand was Heavy Battalion 502 – this formation was sent into the battle piecemeal to shore up endangered sections of the front, its sub-units sent into battle as soon as they arrived. This lead to disaster for the 3rd Company, which had all of its 15 Tigers destroyed or captured after being surrounded by elements of the 42nd Army, the lieutenant in command committing suicide.

It should be noted that the remainder of Heavy Battalion 502 gave invaluable service covering the 18th Army's retreat to the Panther Line, stopping several Soviet attempts to preempt their occupation of this position. They were met by the 12th Panzer Division in early February – this unit had only a single tank battalion, however, equipped with a mix of early and late model Pzkw IIIs and IVs, yet another victim of the Germans choosing to focus on new designs instead of completely reequipping its armored formations with updated versions of the Pzkw IV. Due to its low complement of effective armor, this division was ineffective in its attempts to stop the Soviet advance on the Luga River. The Soviets managed to occupy Luga itself on February 13th, rendering the defense line unviable and forcing the Germans to withdraw still further to the Panther Line, which was anchored at Narva and the Baltic Sea to the north and followed the western edge of Lake Peipus, going through Pskov and then Vitebsk.

It is important to note that the Soviets used large numbers of light and obsolete tanks well into 1944 - here we see a unit of T-70 light tanks on the move near Leningrad. While these tanks could be useful for infantry support, they could also rapidly add to the score of any tank or assault gun commander who caught them out in the open.(RGAKFD)

In this defensive position, the Tigers of the 502nd Heavy Battalion along with some Panthers were able to aid the German infantry and Stug battalions in holding the line until summer, when the massive Soviet Belorussian Offensive outflanked the position, forcing the Germans to abandon it. The Tigers and Panthers had proven very useful here, where the short distances required to travel to different trouble spots did not overly tax their engines. The availability of supply depots and well-appointed repair facilities close at hand was also helpful. This was not World War I, however, and any positional fighting was unlikely to last long. Farther to the south, Panthers and Tigers would be thrown into battle once again in an attempt to rescue their trapped comrades at Korsun.

While of not much use in mobile warfare, the Tiger found its forte in defensive positional fighting.(Bundesarchiv)

It's worth mentioning that the Soviets could have done better against Army Group North if they had major mobile operational units available such as tank or mechanized cavalry corps, with the latter being an especially good fit for this area of operations with its ability to travel cross-country over poor terrain with ease and its reduced need for fuel and other logistical support. Such units would have been able to surround German formations instead of just pushing them back before they could withdraw in good order. Indeed, it became regular Soviet practice later in 1944 to attach a tank corps to every combined arms army in order to take advantage of such opportunities.

CHAPTER 6: THE KORSUN POCKET

Meanwhile, Soviet efforts in the Ukraine had created a salient at Korsun, where German units continued to hold along the Dnepr River. The encirclement battle began on January 24th when Konev's 2nd Ukrainian Front attacked from the south-east. The 4th Guards and 53rd Army rapidly broke through the lines of the defending 389th Infantry Division, allowing the 5th GTA to be committed into the breach to make contact with Vatutin's 1st Ukrainian Front, which was attacking from the north-west. The Germans attempted a counter-attack with the 11th Panzer Division, but like so many of their attacks during this period it failed to achieve meaningful results due to lack of sufficient mass, having only 15 tanks and 15 assault guns remaining in its order of battle. By the 27th, the Soviet Tank Army was close to linking up with Vatutin's forces.

Soviet 85mm AT gun. AT regiments equipped with this and other AT guns were often able to halt attacks by German Panzer Divisions. In addition to static AT weapons, these regiments were often equipped with self-propelled AT guns and mobile obstacle detachments as well.(RGAKFD)

Another German effort to defeat the 5th GTA before this occurred materialized on January 28th, when the Panther equipped 26th Panzer Regiment was committed in an attempt to cut off the Soviet spearhead. After losing 17 Panthers to mechanical failure reaching the start line, the attack went in at 0600 hours. In a display of hubris and overconfidence in their vehicles, the Germans performed no reconnaissance and ran into an ambush by Soviet AT regiments and units of the 18th Tank Corps. A further 28 of the regiment's 61 Panthers were lost, with another 16 lost to breakdowns over the course of combat. The German attack failed and the 18th Tank Corps succeeded in driving off the 11th Panzer Division and reopening the lines of communication with the 5th GTA. With their rear area secure, the 5th GTA was able to seize the German repair and supply depot of Zvenigorodka, meeting up with lend-lease Sherman tanks from the 6th Tank Army on January 18th. The encirclement of 59,000 troops within the Korsun pocket was now complete.

Soviet infantry passing Panthers abandoned after one of the early German counter-attacks against the 5th GTA. (RGAKFD)

Since Hitler would not allow the encircled forces to withdraw, the first of several relief forces began to be organized to reopen land communication with them.

The first effort came from the 47th Panzer Corps on February 1st, which sent the somewhat spent 11th Panzer Division that had only 22 Panthers and 3 Pzkw IVs remaining, along with 13 Stug assault guns. The Soviet rifle divisions that were present hadn't organized their anti-tank defenses yet, however, and the armored vehicles of the relief force were able to drive straight through them. The Germans advanced over 30 kilometers in a day thanks to this disorganization, stopping within striking distance of their encircled comrades whose combined forces were now called 'Group Stemmerman' after the officer in overall command - only the waters of the Shpolka River stood between them and success. Unfortunately, the bridge they had found to cross it collapsed as the Panthers drove over, spilling at least one into the water and halting the advance. To compound the problem, neither the 47th Panzer Corps nor its controlling 8th Army had any ferries or mobile bridges capable of sustaining the Panther's weight.

In the end, the collapse of the bridge was fortuitous – sending a weak panzer division to the rescue of Stemmerman without any support or even a follow-on echelon was foolhardy. The 11th Panzer remained in place while reinforcements were brought up in the form of the 24th Panzer Division, a unit just as poorly off as the 11th, but it was a start. Sure enough, the Soviet 29th Tank Corps and 49th Rifle Corps soon showed up and prevented the 47th Panzer Corps from advancing any further, but had it caught the 11th Panzer Division alone and unsupported, the division would have been annihilated.

The Germans did manufacture an all-terrain logistics vehicle called the Raupenschlepper Ost, specifically for the sorts of environmental conditions met on the Eastern Front. Even with thousands manufactured, there were never enough of them to fully meet the needs of the German mechanized formations. (Bundesarchiv)

The German's next attempt to relieve the pocket was more substantial, involving the forces of the 3rd Panzer Corps. There was a delay in assembling the necessary forces as many of the units participating were already involved in combat elsewhere. Mud was another problem, and another reminder that the Germans would have been better off designing four wheel drive trucks instead of heavy tanks back in 1942. Vehicles like these would have been able to bring up fuel, munitions and other supplies, lack of which would ultimately doom the relief of Group Stemmerman. The battered units that arrived in the assembly area consisted of the 17th and 16th Panzer Divisions, Heavy Battalion 506 and Heavy Panzer Regiment Bake, a unit equipped with both Panthers and Tigers. All told, the assembled units had the strength of one reinforced panzer division, with 48 Panthers, 41 Pzkw IV, 16 Tigers and 21 Stug III. All of them were tired and under supplied, but the attack went in anyway on February 4th. With two infantry divisions in support, the panzer corps was able to advance 19 kilometers on the first day against the 104th Rifle Corps.

By day two, however, the Germans ran into two insurmountable obstacles: the Gniloy Tikich River and the 6th Tank Army, which received the support of an AT brigade in order to halt the advance. Again, the weight of the heavy tanks in the German panzer divisions prevented the use of bridges across the river while the 6th Tank Army with its agile T-34s harried the flanks of the advance. Even when reinforced by the 1st Panzer and 1st SS Panzer Divisions, the 3rd Panzer Corps could go no further, its units out of fuel and ammunition by the end of the second day of the operation with no reliable means of bringing these vital supplies forward.

Reorganizing its forces, the 3rd Panzer Corps tried again on February 11th. Despite the corps' formations being down to 140 tanks, the Germans managed to seize a crossing over the Gniloy Tikich near Lisyanka capable of bearing the weight of their Tigers and Panthers. Lisyanka itself was captured soon after, but then the Germans had to wait another full day for resupply, the fuel trucks once again delayed by mud as well as being constantly attacked by bypassed Soviet units.

By February 13th, 3rd Panzer Corps was able to attack again, but with a battle group from the 1st Panzer Division and the remnants of Heavy Panzer Regiment Bake, which had been reduced to ten Tigers and ten Panthers.

Tigers with infantry escort heading toward the front at Korsun. (Bundesarchiv)

Meanwhile, the 2nd Tank Army had arrived on the scene, sending four tank brigades into the path of the German advance. Bake's regiment and Kampfgruppe Frank from the 1st Panzer Division joined forces to try and break through the positions of the 50th Tank Brigade, but ran into a mixed battle group of T-34s and AT guns. They managed to advance another 12 kilometers, but lost half of their armor in the engagement, leaving Bake with only five Tigers and four Panthers to continue their advance – and they still had a further 10 kilometers to travel to reach their trapped comrades. To add to their difficulties, they were once again out of fuel.

On February 14th, Luftwaffe JU-52 transport aircraft tried dropping fuel drums to the relief force so they could renew their drive to the pocket. While most of these burst on hitting the ground, enough was recovered to get the battle group's vehicles moving again.

As it turned out, no more meaningful advances could be made, the Germans losing another four Tigers and three Panthers for their efforts and once again exhausting their fuel. The Soviets, using the superior mobility of their logistical vehicles and armor, simply kept putting more tanks, AT guns, self-propelled guns and artillery into the path of the Germans.

The Soviets went through their own process of trial and error trying to develop a usable heavy tank. The KV-85 was not one of the more successful designs,being mechanically temperamental and armed with the same 85mm gun as the T-34/85. (RGAKFD)

In the meantime time was running out for the trapped Group Stemmerman, the Soviets having captured the airfield the Germans had been using to keep supplied on February 13th. 3rd Panzer Corps tried another push on February 16th. After losing still more heavy tanks to Soviet KV-85s, the relief force ground another three kilometers forward to try and shorten the distance Stemmerman would have to cross – the breakout would occur that night.

Things started off well, with the 72nd Infantry Division quietly making its way after dark through the lines of the 5th GTA. This respite was brief, however, with the lead tanks of the 5th SS Panzer Division running into an ambush from AT and tank units from the 18th Tank Corps. The 5th Guards Cavalry Corps, alerted to the breakout attempt, attacked along the southern flank of the penetration, hitting many of the support units that were following the German combat formations, causing them to panic. A rout ensued, with all of the evacuating units losing any semblance of cohesion. 3rd Panzer Corps, with no fuel and few tanks, was unable to assist. 35,000 Germans managed to find a way to cross the frigid Gniloi Tikich River, leaving all of their tanks, artillery and other heavy weapons behind – many were without their personal weapons as well.

Soviet Cavalry advancing – unlike more romantic portrayals, it was very rare for them to charge their enemies with sabers drawn. They

instead acted more like dragoons, dismounting and fighting as infantry as soon as their reconnaissance elements made contact with the enemy.(RGAKFD)

Even while receiving priority for replacements, it took the 5th SS Panzer Division six months to once again place itself in the front line. A total of 19,000 German soldiers were killed or captured during the breakout.

In addition to the units lost from Group Stemmerman, the 3rd and 4th Panzer Corps were spent. These formations had permanently lost 236 tanks and assault guns from all purposes, with most of the remaining armor in poor mechanical condition. By the end of February, 171 of the 187 Panthers remaining to the 1st Panzer Army were immobilized due to damage or mechanical problems. Many of these would be abandoned in subsequent retreats – and retreat is what the Germans would have to do. Having expended their armor and logistic reserves at Korsun, there was little left to stop the Soviet tank armies from overrunning the rest of the Ukraine.

The other half of the secret of Soviet sustainable offensive operations - the first half was the T-34, the second half was American Lend-Lease trucks like this one. These trucks began arriving in massive numbers in 1944 and were used to mobilize infantry in the mechanized armies of the Soviet Union as well as sustain them with supplies during their advance. Reliable and with good off-road mobility, trucks like these were a key element in the Allied victory in World War II. (RGAKFD)

German trucks, on the other hand, tended to be much less well suited for inclement weather and terrain. In addition, the Germans used several different models of trucks, many of them captured from enemies. This made spare parts for them very difficult to come by. (Bundesarchiv)

The German failure at Korsun can be traced to two essential factors – lack of mobility in adverse terrain in their combat and logistical formations and lack of reliability in the armored vehicles they were relying on to sustain their combat operations. Finally, the Soviets were demonstrating an increasing ability to overmatch their opponents operationally by being able to sustain breakthroughs and utilize their superior mobility to block German attempts to interfere. The use of formations like mechanized cavalry crops facilitated Soviet operations in adverse terrain and weather and could operate in regions that couldn't be traversed by Axis mechanized units.

CHAPTER 7: ANZIO

Meanwhile, the Western Allies were being introduced to the Panther at Anzio in Italy, with Heavy Battalion 508 with its Tiger tanks also putting in an appearance. The Anzio amphibious landings were Churchill's brainchild and suffered from an under abundance of forces committed and an over abundance of objectives. The amount of reserves the Germans could bring to bear was also underestimated, as was the speed with which they could be introduced.

While the allied advance after their initial, nearly unopposed landing could be considered cautious, it should be kept in mind that this slow pace was due to their desire to consolidate their bridgehead, a lesson learned from the near disastrous Salerno landings months before.

The confusion over objectives came as a direct result of Churchill's Mediterranean strategy, which sought a post-war political situation in this region that favored British interests, most notably security for the Suez Canal and the route to India. In the short term, the allies wanted to use the Anzio landings as a way to bypass the Gustav Line to the south, which had withstood all attempts to breach it, seize Rome at the earliest opportunity and cut off the German 10th Army manning the line by seizing the Alban Hills.

The single corps that was initially landed wasn't capable of taking even one of these objectives, let alone both of them, and it was this miscalculation that led to the near disaster in the form of German counter-attacks meant to drive the allied force back into the sea. This was a very real possibility as the German reinforcements soon outnumbered the American and British forces and included a battalion each of Tiger and Panther tanks, both initially attached to the 29th Panzergrenadier Division. These attacks may have succeeded against a less organized enemy – as it was, they were defeated by a stubborn allied combined arms defense with an emphasis on artillery amply supplied by superior allied logistics. The Tigers and Panthers were part of a menagerie of German armor sent to contain and eliminate the allied presence, none of which was well suited to the muddy terrain in the area. The Panthers saw little use except as long range gunnery platforms, something their high velocity 75mm guns were ill suited for since their targets were primarily infantry. The Tigers of Heavy Battalion 508 had a more direct role, however, illustrated here in their role in Operation Seitensprung where the 2nd Company (attached to Panzer Division Herman Goering) participated in the final attempt to eliminate the allied bridgehead.

The Tigers had seen combat in the area before, operating out of a strongly built agricultural village called 'The Factory' by allied units due to the resemblance of its church's steeple to a smoke stack. The battalion's initial deployment was problematic since its Tigers were off-loaded 200 kilometers away from the battlefield with most of them breaking down on the way, one of them overheating to the point of the engine catching fire and destroying the tank. Because of this, when the unit was committed to combat on February 16th, it did so far understrength, a problem that would continue to plague the heavies for the rest of the war due to having to be dropped off far from the action because of allied air interdiction. Restricted to primary roads because of the muddy terrain, the few Tigers available on any given day had little impact on the course of the fighting.

When a more concentrated force was assembled for Seitensprung on the 29th, the results were also less than decisive. Like before, muddy terrain restricted the Tigers to a narrow axis of attack along the few paved roads. After first being halted by mines, the column then came under fire by artillery. Engineers were called up to clear them, but could not do so in the face of the continued barrage as one by one, the Tigers were knocked out or immobilized by mines, artillery or direct fire from American M-10 tank destroyers. Ironically, the Germans were assisted in their recovery of their damaged vehicles by using captured Sherman tanks with their turrets removed and pressed into service as towing vehicles. When night fell, the remnants of the heavy tank company retreated back to their own lines, having accomplished little.

Both the Tiger and Panther had overlapping wheels to provide a smoother ride – these proved to be problematic in service as they would often become clogged with mud, ice or other debris and were difficult to remove. (Bundesarchiv)

The role of Tiger and Panther tanks during the remainder of their service in Italy tells a similar story. During the German attempt May 21st - 25th to stop the link-up of allied forces at Anzio with those breaking through the Gustav Line. During this attack, the Tigers of Heavy Battalion 508 engaged allied infantry and tanks, destroying 15 Shermans and advancing three kilometers, briefly delaying the occupation of Rome but losing a total of 22 Tigers though enemy action and mechanical breakdowns. Recovery efforts utilizing Tigers failed due to transmission problems with the tanks attempting the tow, with one stranded vehicle being recovered only after the same two captured Sherman tanks mentioned previously were sent from Rome to get it out of trouble. The remaining Tigers had to be destroyed by their own crews.

Despite the lackluster performance of Heavy Battalion 508, another battalion, number 504, was sent to join it in June. Not surprisingly, it encountered all of the same problems of its sister unit, with 28 of its 45 tanks lost during a ten day period in combat with the US 1st Armored Division. Few of these tanks were lost in combat – most slid off mountain roads, collapsed bridges that couldn't bear their weight (stranding any units unfortunate enough to be behind them) or simply broke down, many times in mountain passes that were then effectively blocked until the tanks could be removed. Meanwhile, the 1st Armored Division was able to maintain an advance of six kilometers a day in the same terrain, aided by the efforts of the 16th Armored Engineer Battalion which swiftly cleared minefields, built new bridges or repaired damaged ones and bulldozed roads to clear them of rubble and other obstacles.

The two Tiger battalions assigned to Italy did better after the front stabilized when German forces reached the fortified Gothic Line. After losing still more Tigers during the withdrawal (Heavy Battalion 508 losing 11), the remaining vehicles were spread out to locations where they could have wide fields of fire and put in restrictive terrain where they would be difficult to outflank. Even though this made supplying and maintaining communications with them difficult, it minimized breakdowns since they didn't have to move much.

By 1945, Heavy Battalion 508 had ceased to exist, handing over its remaining 15 tanks in various states of repair to the 504th Battalion. When the 504th surrendered to the allies, it had no operational tanks remaining. Overall, the heavy tank battalions in Italy claimed about 200 tanks destroyed – in reality, given over claiming and taking allied tanks that had only received minor damage after being hit into account, the actual number is probably closer to 100 or less. The heavy battalions themselves lost a total of 157 Tigers to all purposes.

The only substantial contribution they made during the campaign was helping to delay allied forces along the Gothic Line. As mentioned before, similar missions could have been undertaken by more reliable vehicles which would have had the added benefit of putting less of a strain on the already threadbare German logistical system which had to supply Tigers with their own unique set of spare parts, which they always needed in quantity.

American soldiers inspect a Panther abandoned near Anzio. (NARA)

Information about the use of Panthers in Italy is harder to come across, but from what information is available, it seems that there were never more than 76 stationed in Italy, assigned to the 1st battalion of Panzer Regiment 4. Having many of the same mobility problems as the Tiger in the mountainous terrain, only a fraction of this number were ever available for operations at a given time after their introduction at Anzio. By May 19th, there were still 62 available, this relatively high number being explained by the unit being deployed in nearly static positions, providing long range bombardment of the allied units being besieged there.

When Panzer Regiment 4 began to withdraw later in the month, there were still 48 Panthers, but only 13 were operational, with many of the disabled vehicles having to be left behind – by June 6th, there only 6 runners left with others under repair, over 30 having been lost during the retreat.

Over the course of the remainder of the war in Italy, there were in general never more than a dozen Panthers in working order at any given time, having a negligible effect on the outcome of the campaign. While the German heavies had a small part to play in the defense of Italy, they would have a larger role in the drama unfolding in the wider war, in locations widely separated by in both distance and circumstance – Normandy and Romania. In the latter region, the German heavies participated in what would be a meaningful operational success.

CHAPTER 8: ROMANIAN INTERLUDE

The First Battle of Jassy in Romania pitted a combined German and Romanian force against an invading Soviet army. The situation was ideal for the German mobile formations taking part, which included the Gross Deutschland, 24th Panzer and 3rd SS Panzer Divisions. These formations had been recently reinforced, were close to their logistical bases and repair depots and were well supported by infantry, AT guns, artillery and even strong Luftwaffe support – a rarity in this period of the war.

The small battlefield and the Soviet's failure to utilize their superior mobility to attack in more peripheral areas meant that breakdowns were rare, with Tiger and Panther formations able to field larger contingents than average. The Soviets, in comparison, had just advanced hundreds of kilometers after their victory in the Ukraine. Their vehicles were in poor repair, their infantry exhausted and all of their formations were badly in need of replacements. Their logistical state could at best be called precarious and air cover insufficient due to friendly airfields being left behind due to the ground forces' rapid advance. In these circumstances, the German heavies contributed decisively to the Soviet repulse, supporting the infantry divisions and leading counter-attacks that kept the Soviets off-balance, unable to assemble and coordinate the forces necessary in the months of April through May 1944.

Bundesarchiv, Bild 101I-244-2321-34
Foto: Waidelich | 1944 August - September

A Panther crew looking for targets in Romania. (Bundesarchiv)

However, even this well organized defense was ineffective against a properly orchestrated offense as the Second Jassy-Kishinev Offensive in late August proved. On such a vast front, the mobile reserves that made the active German defense of Romania possible were in short supply, and many of them were drawn off to the north earlier in the summer in an attempt to deal with the catastrophe unfolding near Minsk during Operation Bagration. These were not entirely absent, however, given the presence of the 13th Panzer Division, 10th Panzergrenadier and 1st Romanian Armored Division. The 13th Panzer had been recently reformed and included the latest model Panther tanks, the G model. Despite this, these formations were consumed in ineffective counter-attacks, with the 13th Panzer Division being combat ineffective after the first day during its run-in with the Soviet 66th Rifle Corps. The Soviet offensive proceeded to overrun much of Romania, with the German 6th Army and parts of the 8th encircled and destroyed. On August 23rd, a coup led by Romanian King Michael overthrew the Axis friendly dictator Jon Atonescu. The German retreat would continue into Hungary where the fighting would drag on for several more months.

CHAPTER 9: NORMANDY

Meanwhile, the German heavy armor battalions were engaged on a new front when the Western Allies invaded Normandy. All of these were initially deployed against the Commonwealth forces in the Caen sector since the Germans feared a breakout here the most, while lighter units screened the US advance through the hedgerows.

The most famous encounter between the German heavies and the Commonwealth forces is the tale of Michael Wittmann in the Battle of Villers-Bocage. Given the relatively compact and static nature of the battlefield, Normandy was an ideal location for deployment of the Tiger and Panther. Once they arrived in the area of battle from the interior, the distances that needed to be traveled were short, which minimized breakdowns. If a vehicle did suffer a mechanical failure or was damaged, there was a good chance it could be recovered and repaired behind the stable battle lines.

But first they had to arrive, which did present some difficulty as the closest the German mobile units dared to travel via train was Paris. Heavy Battalion 101 arrived in Normandy on the evening of June 12th, but only had six Tigers available, with most of the unit's tanks having broken down along the way, strung out on the road between Paris and Normandy. More were enroute, but would not arrive until the 15th. Despite all of the advantages offered by the nature of the Normandy campaign, in no instance were the Tiger battalions able to operate at anywhere near full strength – usually they were committed in small parcels numbering no more than six or twelve tanks while the rest of their units languished in repair depots.

Nevertheless, even these small numbers could make a difference on the operational level if given proper support, easing operations that would have been more difficult without their presence.

Bundesarchiv, Bild 101I-738-0276-10A
Foto: Grimm, Arthur | Juni 1944

Tigers near Villers-Bocage (Bundesarchiv)

Villers-Bocage is the encounter even those with little knowledge of World War II have heard about, and indeed it may have been this one engagement that cemented the fame of the Tiger firmly in the place it holds in the military ethos. In reality, the Tiger's role in general and Wittmann's in particular tends to be overemphasized. Accounts tend to portray Wittmann stopping the advance of the entire British 7th Armored Division by himself – in fact, Wittmann had the assistance of the nearby Panzer Lehr Division as well as the reconnaissance battalion of the approaching 2nd Panzer Division.

The forces employed by the Allies included a brigade sized task force made of elements of the 7th Armored Division, with much of the latter engaged with the nearby Panzer Lehr Division. First by himself and then with the assistance of other vehicles from his company, Wittmann managed to wreak havoc with the advance group of the British brigade, causing the British force to halt and then retreat into the village. Wittmann pursued them, and his tank was promptly knocked out by a 57mm AT gun. A wiser course of action would have been to withdraw and wait for infantry support before proceeding into the close confines of the village. Wittmann and his crew escaped from their damaged vehicle, making their way on foot to the Panzer Lehr Division HQ. After informing them of the situation, a second attack was organized against the village, but this was again repulsed by the British.

The presence of elements from two German armored formations, especially the 2nd Panzer Division, convinced the British that their position was exposed (which in fact it was) and caused them to withdraw. While this was a clear British defeat, the fact remains that the force they committed would not have been able to overcome the German reinforcements that were being introduced into the area. Indeed, a deep advance into the German rear areas may well have been cut off by the 2nd Panzer Division, resulting in an even greater debacle. The wisdom of the British retreat is further illustrated by the continued heavy attacks on June 14th against the 22nd Armored Brigade by the Panzer Lehr Division in a position the defenders called "The Island" due to its exposed nature, being surrounded by German forces on three sides. Though the German attack was once again turned back, the British decided to withdraw their brigade still further to make their lines more defensible, thus ending their ill-fated foray into German territory.

Even through the German effort was crowned with success, fundamental errors were made regarding the use of tanks in built up areas without adequate infantry support. While Wittmann's attack did stop the advance of the 22nd Armored Brigade, the damage inflicted was minor and his pursuit of the enemy into the village itself was foolhardy, putting his life and the rest of his experienced crew at needless risk. The follow-on German attack into the village repeated this mistake, with much valuable German armor being lost. A wiser course of action would have been to make a fighting withdrawal, continuing to inflict losses on the British as they advanced. This would have brought the Tiger's advantage of strong armor and effective gun to the fore, and would have lured the 22nd Armored Brigade further forward, allowing arriving German reinforcements to cut it off and destroy it.

The British answer to the German heavies – the 17 Pounder AT gun. (Author's collection)

Throughout the rest of the fighting in Normandy, the Tigers continued to score some successes of operational note. These successes always came at a cost, however, especially when the heavy vehicles were used in the attack given the Commonwealth usage of the 17 Pounder AT gun, a very high velocity 76mm weapon capable of destroying any German tank at Normandy with the exception of the very rare Tiger II. This weapon was available in both towed and self-propelled mounts, the latter in the form of the Firefly and Achilles tank destroyers. While these vehicles had lighter armor than the German heavies, they were far more reliable and easier to produce, guaranteeing that there would always be more of them present on battlefield. In a battle between 17 Pounder equipped units and German heavy tanks, victory would go to who spotted their enemy and fired first.

During the Normandy fighting, Heavy Battalion 102 assisted the 9th and 10th SS Panzer Divisions in keeping Commonwealth forces off of Hill 112 until forced to withdraw due to threats to their flanks caused by Operations Bluecoat and Cobra, although this did not happen until August 4th. Possession of this hill was extremely important to both sides, as it allowed for excellent observation of much of the ground around Caen. While the Germans had held for a considerable amount of time, it was at a heavy cost due to the hill's exposure to conventional and naval artillery.

The destruction of Worthington Force, a Canadian battle group that participated in Operation Totalize, was another noteworthy use of the Tiger in Normandy. During this engagement Worthington Force, consisting of the Columbia Regiment of Sherman tanks from the 4th Canadian Armored Brigade and two infantry companies mounted in Priest and Kangaroo Armored Personnel Carriers, had gotten hopelessly lost during its night attack when it infiltrated deeply into German positions. It was soon detected and nearly annihilated by counter-attacking elements of Heavy Battalion 101 with eight operational Tigers and 24 Pzkw IV and Panther tanks from the 12th SS Panzergrenadier Division. After surrounding the Canadians, the Germans tore into them with long range fire the lighter Sherman tanks could not effectively reply to – very few escaped back to Commonwealth lines.

Knocked out Panther tank in Normandy. While this model of Panther was more reliable than earlier versions, it still suffered from abysmal availability rates – and by 1944, the Allied powers had developed counter-weapons for the rare occasions when they were encountered. (NARA)

Michael Wittmann, meanwhile, met his end as part of a counter-attack against Operation Totalize that also included elements of the 12SS Panzer Division. The counter-attack, which was intended to stop or at least delay the start of the second part of the Commonwealth offensive, consisted of 500 dismounted panzergrenadiers, ten JagdPanzer IV tank destroyers, a mixed group of 20 Panzer IV and Panthers and the four Tiger tanks that remained to Wittmann's company. This battle group was divided, with most of the tanks and infantry forming a attack from the east, with Wittman's Tigers approaching from the west. The Germans were unaware that Sherman tanks with their accompanying Firefly tank destroyers from the Northamptonshire Yeomanry were occupying an orchard that they would be passing – during the ensuing ambush, all four Tigers were rapidly destroyed with no to loss to the Yeomanry. The turret of Wittmann's tank was blown off during the exchange, with him and his entire crew killed.

It is always more efficient to use an existing vehicle to fill a battlefield need. The Firefly is an excellent example of this, utilizing a Sherman chassis to create a very effective tank destroyer. (Author's collection)

The Tiger II was also deployed for the first time against Commonwealth forces in Normandy. Meant as an improvement over the Tiger Mk 1, it had a better gun in the form of an 88mm L/71 and enhanced, sloped armor up to 185mm thick. The gun could easily destroy any allied vehicle encountered in the east or west and the armor along the frontal arc was impervious to all but the heaviest allied AT weapons. But as mentioned before, all of these deadly qualities are useless if the vehicle can't be produced in meaningful quantity or get to the battlefield because of mobility and reliability issues. Weighing in at nearly 70 tons, the Tiger II was even more of a gas hog than its predecessors, needed more maintenance, could use fewer bridges, broke down more often and when it did, was nearly impossible to move. It also added yet another vehicle to the German order of battle, complicating the logistical situation still further with yet another unique set of spares.

The Tiger II – while physically impressive, it was an extremely impractical use of Germany's dwindling resources. (Author's collection)

That the Germans not only continued to manufacture their previous heavy designs but doubled down on the idea and started production on another behemoth taxes the imagination as to why they thought this would improve the military situation in their favor. Reading after action reports by German crews helps clear this up a little – they seem to be blind as to the shortcomings of their vehicles, with most of the reports filled with praise as to its excellent gun and armor. They'll often mention how few of them there are, but never reflect as to why that's the case. While such attitudes might be excused from enlisted personnel and junior grade officers, people in higher echelons of command with a larger overall picture of accelerating mission failure from 1943 on should have known better.

That being said, the Tiger II did participate in several engagements, but for the above mentioned reasons, did not have a decisive impact on operations. A good example of its employment in Normandy concerns a counter-attack by elements of Heavy Battalion 503 during Operation Bluecoat, a Commonwealth endeavor to draw off German reserves and protect the flank of the American attempt to break through the hedgerows known as Operation Cobra. Several Tigers were committed to counter-attack the incursion near Vire, but were halted by a rapidly hardening defense after some initial progress. The Germans lost several of their attacking vehicles and at no point was either allied offensive endangered.

The Cromwell -while not as imposing as the Panther or Tiger, it was eminently more well-rounded and useful. (Author's collection)

The Commonwealth took losses as well, of course, losing many Cromwell tanks knocked out or damaged. These of course were much more easily replaced, and many of those that were knocked out could be rapidly repaired and put back into action. Many accounts of the fighting in Normandy use the German method of counting losses, and this provides a distorted picture of armor losses in the west much as it does in the east. Every tank the Germans hit was counted as a permanent loss, while their own vehicles were counted as losses only if they burned, exploded or were abandoned. Their own damaged tanks weren't counted as losses at all, even though there was a very good chance that they would never see action again, either spending their remaining days languishing in a repair depot or being cannibalized for spare parts.

For example, the British 8th Corps is often cited as having lost 314 tanks during Operation Goodwood between July 18th – 20th. In fact, only 130 had been completely destroyed, with many of the rest soon back in action after being quickly repaired. German permanent losses were 75 tanks, but it should also be kept in mind that many Commonwealth losses would have be caused by towed AT guns and infantry light AT weapons such as the Panzerfaust since the Germans were defending.

The Panther was also present at Normandy, but if anything its presence had even less operational significance than the Tiger. It was mostly used against the Commonwealth forces, which as mentioned had an excellent anti-tank capability in their 17 Pounder AT gun. The Panther's armor was not proof against this weapon, and every Commonwealth tank troop had one in the form of a Sherman Firefly. Many more examples were attached to higher headquarters to be deployed as needed, and these units could respond rapidly to any attack by German heavy units. Victory would go to whomever spotted their enemy first and got off the first shot.

Unlike the Tiger, the Panther was also deployed against the Americans, but the close range nature of the hedgerow fighting nullified any advantages the Panther might have had. An example would be the ambush of a column of Panthers belonging to the Panzer Lehr Division near the village of Le Desert by M-10s of the 899th Tank Destroyer Battalion. All ten Panthers were rapidly destroyed without loss to the attackers.

The M-10 proved time and again that it was perfectly capable of knocking out the German heavies. Friends included for scale. (Author's collection)

Despite the advantageous nature of the fighting in Normandy to German heavy tank units in that it was mostly positional until the Allied breakthrough, the Germans could not turn the presence of these units into a decisive edge due the presence of hedgerow terrain in the American sector, the presence of advanced AT weapons in the Commonwealth sector, continuous allied attacks that kept German reserves off-balance and unable to concentrate for a decisive attack against the bridgehead and finally, the continued unreliability of the heavy tank units that kept large numbers of them absent for extended periods. Once the breakthrough occurred, these immobile vehicles would be left behind during the German retreat, just as they had been in the east.

Normandy had been the last chance for Germany's heavy tank units to prove themselves. The war still had several more months to play out, but never again would they have an opportunity to decisively smash a major allied offensive and possibly turn the course of the war.

Abandoned Panzers at Normandy (UKNA)

CHAPTER 10: THE TIGER AND PANTHER - CONCLUSIONS

The prospect of making a nearly invulnerable and deadly but still usable tank was briefly realized by the Soviet Union in 1941 – 42 with the T-34. Accelerating technology in anti-tank weaponry changed this, but to their credit the Soviets focused on producing the same vehicle but improving their logistics, tactics and communications, which in the long run turned out to be a far superior force multiplier. They also concentrated on making a large number of a few different designs, simplifying both their production and logistics. The tactics they designed relied upon having hundreds of reliable tanks available at any given time for their operations – while these were not the best protected and best armed tanks in the field, there were always enough for infantry support duties (independent tank regiments) and exploitation of breakthroughs (tank armies). Their success speaks for itself, delivering actual operational victories instead of the occasional thrilling tactical narrative, that, while making for fun reading for German tank enthusiasts, were a poor substitute for the support the Heer so desperately needed.

The Americans followed a similar strategy to the Soviets, focusing on the reliable and easily produced Sherman. This vehicle was produced in such quantity that the Americans were not only able to lavishly equip their own forces, but were able to supply their allies as well, with M4s equipping entire Commonwealth and Soviet tank brigades by 1944. The Americans didn't focus on maneuver as much as the Soviet Union, but this was natural given smaller overall size of the theater they were engaged in. Their firepower intensive tactics were more than adequate given their strength in logistics, and they could certainly maneuver if given the space and opportunity as the Normandy breakout and the surrounding of the Rhur in 1945 demonstrates.

The M4 was a well-balanced armored vehicle that served in every major Allied army, contributing decisively to their overall success. (Author's collection)

The Germans, meanwhile, tried to make invulnerable, heavily armed tanks, not realizing that this approach was not viable given the limitations of the technology of the day. They went ahead anyway despite early indications of their heavy tanks' shortcomings, applying the same 'we'll figure it out as we go' mentality to their vehicle production as they did to their military operations. Faith in their National-Socialist engineers also played a role in this – any vehicles they developed had to be by definition the best. The propaganda surrounding these vehicles was so effective, it not only convinced the Germans but later generations of foreigners as well as to the superiority of their designs.

Even the prospect of using such tanks in positional warfare like the fighting around Leningrad, the Panther Line, Italy and Normandy was no longer necessarily a guarantee of good results due to the existence of allied AT weapons capable of destroying them. Attempts to stay ahead of these increasingly deadly allied counter-weapons merely resulted in utterly unreliable vehicles like the Tiger II, Jagdtiger and Sturmtiger. While there were few allied weapons capable of destroying these vehicles, they were largely unnecessary as the super heavy tanks tended to take themselves out of the picture by breaking down well before they reached the battlefield.

The German shortage of tanks from 1943 to the end of the war is constantly commented upon, and is usually put forward as some sort of unfair allied advantage that Axis forces must somehow overcome. The fact that this came about due to decisions deliberately made by the upper echelons of the German government to produce fewer 'high quality' vehicles is seldom if ever mentioned.

The popularity of the German heavy tanks with the soldiers who used them and the troops they supported comes as no surprise. They could be very effective from the limited tactical point of view of the average soldier on the ground, who didn't notice or care how much fuel the vehicles were consuming or how difficult they were to maintain. They also kept their crews relatively safe from enemy fire, turning aside shot that would have destroyed vehicles with less armor, while their main armament could overcome the tanks of their opponents from great distances. If the crew had to destroy the vehicle after it broke down, they could always get another – if they had to sit out a battle because their tank was being repaired, there was always the next one or the one after. Hopefully, some enterprising infantry commander wouldn't dragoon them into his company as replacements while they were waiting. It was the after-action reports written by these men, in addition to the hype by the National-Socialist press, that was read by later generations of tank enthusiasts and the myth of German tank supremacy was born.

Another factor was how these weapons were seen from the allied perspective. Their sheer physical size could be terrifying – some had their vehicles destroyed by them, or watched as their shots bounced off their heavy armor. That this didn't happen too terribly often is lost in the reading of these exiting encounters, and this feeds into the myth as well. The German heavies left such a large psychological impression that soon every one of their tanks that was encountered became a 'Tiger', even if none were in the area. While their mystique did have a power all its own, one can't help but think that this was a poor substitute for actual combat capability in German units.

After Normandy, the rest of the war was a downward spiral for the Wehrmacht. The 'industrial miracle' of 1944 by German industry was largely an illusion. While large quantities of vehicles were produced, this was done at the expense of replacement engines, spare parts and logistical support vehicles. The Soviet Union had done a similar exercise in the 1930s – while the large production numbers may have been of significant propaganda value, the formations created tended to be brittle, lasting only for one battle or two before disintegrating.

There was also little fuel for the gas hungry heavy tanks due to Western Allied bombing of refineries in addition to the Soviet seizure of the main German source of fuel in Ploesti, Romania. Even the fuel that could be produced had a difficult time making its way forward through increasingly effective air interdiction.

Years of attrition and defeat in the east had mortally wounded the German army – the disaster in Normandy and the destruction of Army Group Center in Belorussia effectively destroyed it. Panthers were massacred at Arracourt in September 1944 by supposedly inferior Sherman tanks and M-10 and M-16 tank destroyers. Redeploying their armor in the east to try and seal the gap created by the destruction of Army Group Center, the Germans left Romania vulnerable – the Soviets took advantage of this absence and seized the country, destroying the German 6th Army in the process.

The rest of the war was pointless offensives by the Wehrmacht and the holding of temporary lines as the Nazis threw less and less capable formations in front of the increasingly professional allied armies – these efforts could at best delay defeat in the amount of time it took for these hollow divisions to be destroyed. Some will point to tactical narratives in 1944 – 45 where the German heavy tanks were victorious to try and maintain that these vehicles were superior, but these incidents had no more outcome on the fighting than similar incidents with heavy allied tanks at Stonne and Arras in France in 1940.

The Panther has the reputation of being the 'best tank' of World War II. Steven Zaloga (whose work I highly admire) even goes so far as to say that it could be considered to be the world's first main battle tank (MBT). As for the first claim, the best one can say is that it might look that way on paper. What's never included in the detailed schematics of these vehicles is how often they were actually present on the battlefield, and how often their superior armor and armament translated into more positive outcomes for the battles in which they were present. As for Zaloga's claim, the T-34 is the first tank that qualifies for this honor, and indeed could perform more of the battlefield tasks of an MBT (infantry support, exploitation of breakthroughs and anti-tank assignments) better than the Panther even though it was an earlier design. The Panther was superior in its anti-tank capability – indeed, it could be considered more of a heavy tank destroyer than a main battle tank given is capabilities. Far from being the war's best tank, it wasn't even the best tank destroyer. While it may have had better armor protection than the British Firefly, the American M-36 or the Soviet SU-100, what really mattered was which vehicle got the first hit in an exchange of fire – in this circumstance, the Panther's extra armor was just added weight. The allied vehicles were also more likely to be encountered in quantities that would make a difference, being more reliable and backed by a superior logistics system that kept them plentifully supplied with fuel and ammunition.

Finally, it should be kept in mind that the design of the Panther wasn't even original – it was an attempt to copy the Soviet T-34 and it failed to do even this. The supposed improvements to the Soviet's design only served to make the vehicle difficult to produce and operate – there's nothing particularly innovative about weighing a design down with so much armor that its transmission continually fails and immobilizes the tank.

Since armor enthusiasts tend to be more aware of the Tiger's shortcomings, it's seldom listed as the Second World War's best tank – it was a useful idea, but poorly executed. Its main armament was very effective in terms of both HE and AT performance – it was also highly resistant to the most common Allied AT weapons, even from the side. However, it suffered from the same problems as the Panther in terms of its mechanical complexity. Despite this, it was able to utilize its advantages to good effect when the front was relatively static or other situations where it didn't have to travel far.

CHAPTER 11: ALLIED VEHICLES

Western allied heavy tanks went through a slow development process, with both the British and the Americans coming up with valid designs by 1944. The British fielded the Churchill MKVII and VIII, vehicles that combined good reliability and armor capable of resisting the most commonly encountered German AT weapons – most notably the 75mm Pak 40. The MKVII had a useful 75mm gun that was capable of engaging soft targets like infantry and artillery successfully as well as the German vehicles it was most likely to fight, the Stug III and Pzkw IV. The MkVII was a specialist close support version with a 95mm howitzer capable of taking out heavier field fortifications. Several engineering variants were produced as well, each of which were invaluable in improving the mobility of Commonwealth armored formations and overcoming the most common obstacle to the Western Allied advance through Europe – infantry and artillery strongpoints, which the Churchill AVRE and Crocodile variants were able to assist their hard pressed infantry in overcoming. The vehicle's major shortcoming was its slow speed, which mostly regulated its service to where the advance was proceeding at a more sluggish pace, like Normandy, the Reichswald or a siege of one of the German held Channel ports. Reliability in the MKVII version and beyond ensured that it was always present in sufficient numbers to give adequate support and provide a positive boost to operational outcomes.

The Churchill AVRE – in addition to its 290mm Petard Mortar, it carried a team of engineers equipped with a variety of different explosives that could be used to remove any obstacle that was blocking a Commonwealth advance. (Author's collection)

The Americans initially had the most practical response to their need for a heavy tank – they modified their M4 medium tank design, creating the M4A3E2 Assault Tank. It should be noted that this design wasn't meant to be a 'breakthrough' tank, but a column leader capable of shrugging off hits from the most common German AT weapons – it was even resistant to the 88mm Pak 36 and the Panther's 75mm L/70. It was armed only with the 75 and 76mm guns found on other Sherman designs instead of the potentially more useful 105mm, which would have been much more effective against soft targets. Even though only 250 were made, they were so popular that many armored units made their own versions in their field workshops. While the extra armor made the E2 a little slower, it didn't noticeably reduce its reliability.

Being a modification of an already existing, proven vehicle, the M4A3E2 wins the prize for being the most practical heavy tank design. Sadly, not many were made, and most that were were assigned to independent tank battalions supporting U.S. infantry divisions. (NARA)

For the sake of completion, the American T26 Pershing should also be included. While the tank's heavy armor made it more resistant to German AT weapons, it did not stand up well to the tanks it was designed to defeat – like so many armor engagements, success in these encounters depended upon who spotted their enemy and engaged first. The T26's main gun had both good HE and AT performance – however, the tank was much more mechanically unreliable than the M4 and its notable that in the Korean War that the Sherman was the preferred tank due to its superior mobility. The Soviet Union ended up having the best heavy tank program of any of the World War II belligerents after something of rough start. The KV-1 and KV-2, like the Tiger and Panther, were good for the occasional spectacular tactical victory that could be exploited for propaganda purposes, but like their later German counterparts on the whole were operational liabilities due to their weight and mechanical unreliability.

A KV-1 at the Bovington Museum with friend M. Conyers provided for scale. To their credit, when the Soviets found a design unreliable, they stopped production and tried a different solution. (Author's collection)

To their credit, the Soviets improved their vehicle by creating the KV-1S. This tank was lighter by way of sacrificing armor protection, and it's improved transmission meant it spent much less time in the repair shop and also made it easier to drive. Its turret had a better layout and was smaller (saving more weight), had a faster traverse and a commander's cupola, all of which made it easier for it to spot enemies and get the all important first shot off in an engagement. When the models that proceeded it proved to be unworkable, production of them ceased, although the chassis of the KV-1 continued to soldier on in the form of the SU-152 – a useful infantry support vehicle with a secondary anti-tank capability.

The KV-1S was later further modified by replacing the turret with that of the vehicle that represented the future of Soviet heavy tank design, the JS-85. Called the KV-85, this tank had the armor to stand up to all but the heaviest German AT weapons, while its 85mm gun could threaten even the German heavy tanks in most circumstances. The 85mm HE round also had superior performance to the 76mm gun carried by the earlier T-34 as well as any German medium tank, the Panther and the Stug III. The KV-85's mechanical shortcomings doomed the design to a short production run and the vehicle along with the earlier KV-1S began to be replaced in Soviet heavy tank regiments with the JS series.

The JS-85 was the first vehicle produced, but since its main armament was the same as the T-34/85 medium tank, it was soon replaced by the JS-2 with a 122mm gun. A 100mm gun with better

AT performance had been available, but the 122mm was chosen due a number of considerations – first, the 122mm was a very common weapon in the Soviet arsenal, so spare parts and ammunition would be easy to come by. Second, the 122mm had superior HE performance while still retaining a significant AT capability. Since the Soviets envisioned the vehicle as being primarily for breaking through German fortified lines manned mostly by infantry and AT guns, the chosen gun was a logical choice. The drawback of this weapon was its low rate of fire (thanks to it using ammunition that came in two parts that needed to be assembled before firing) and the fact that the relatively compact vehicle could only hold 28 rounds of ammunition.

The JS-2 proved to be the best heavy tank design of WWII due to its ability to incorporate superior firepower and armor protection without sacrificing mobility or reliability. (RGAKFD)

Its armored protection from the front was proof against the more commonly encountered German AT weapons and could often deflect even 88mm and 75mm L/70 rounds carried by the Panther and JgPz IV/70 tank destroyers. The side armor provided protection against the 75mm Pak 40, which was by far the most common German AT weapon in 1944, the vehicular version arming the later model PzkwIV, Stug III and IV, JgPz IV, Marder and Hetzer vehicles. Greater armor protection was of course possible, but since this would have compromised the JS-2's mobility and reliability, the Soviets chose not to do this.

The JS-2's small size and well placed, sloped armor stood it in good stead with German heavy tank designs. At 46 tons, it weighed about the same as the Panther despite having far superior all-around armor protection, and of course the 122mm main armament could effectively engage a variety of targets, unlike the Panther's 75mm which was optimized for destroying tanks. Meanwhile, the Tiger IE weighed 57 tons and the Tiger II an unwieldy 70 tons.

The JS-2 also had the other common Soviet virtues of being easier to manufacture and maintain. Ease of manufacture meant large numbers could be fielded in independent tank regiments that could be attached to infantry armies whenever breakthrough operations were about to commence – starting in late 1944, the regiments were expanded to brigades that concentrated the firepower and shock value of these vehicles even more. Tank armies were also given a regiment of JS-2s to use as the situation warranted.

The chassis of the JS-2 was produced in such quantity that it was also utilized for the JSU-122 and

JSU-152 assault guns. These turretless vehicles were often used in a similar fashion to the JS tanks, as overwatch vehicles during an advance over open ground or as an accompaniment to infantry assault teams operating in an urban environment.

Given its strengths, the JS-2 was the best heavy tank of the war – but it wasn't the best tank overall. There is a tie for this honor between the M4A3 Sherman and the T-34/85 with the advantage going to the latter due its wider tracks that improved its off-road mobility over soft ground (although this was later addressed in the M4 in late 1944 with a field modification that attached extenders to the side of the tracks) and the superior performance of its 85mm gun – unlike the later Sherman with its 76mm gun, the upgraded T-34 did not sacrifice HE capability for improved anti-tank performance.

The T-34/85, seen here at the Flying Heritage and Armor Museum in Everett, Washington. (Author's collection)

The Sherman gets an honorable mention for its sheer utility. Used by every allied army, it saw action in every major theater where World War II was fought and performed superbly in every one of them, giving the allies a decisive edge in operational mobility and massed firepower. Manufactured only in the United States, the tanks themselves, as well as the spare parts and ammunition that sustained them were shipped overseas to the fronts that required them. Its mechanical simplicity and compact size meant that many could be manufactured and easily shipped, while its reliability meant that not only would more be present on the front line, less space in freighters would be required for spares and could instead be used for other purposes. An allied victory may still have been possible without the M4, but it would have taken much longer and been far more costly. While the Panther and Tiger could produce isolated tactical victories that may or may not have been relevant to the larger battle being fought, the M4 was an instrument of global operational and strategic success.

In addition to its other fine qualities, the Sherman lent itself to several different types of useful modification, including specialist engineering designs and this up-armored version seen serving with the U.S. 2nd Armored Division in Germany. This particular upgrade was very common with this unit, and was done in field workshops utilizing armor plate recovered from destroyed vehicles. (NARA)

In the end, the decision to produce the Tiger and Panther tanks, and to continue to produce them even after they had proven to be ineffective, was an enormous mistake that made an allied victory easier than it would have been otherwise. Instead of producing these two behemoths, Germany should have followed the example of Soviet and American industry, focusing on one key design that would be produced throughout the war that had the capability to be modified as circumstances dictated, that was mechanically reliable, fuel efficient and easy to produce and repair.

Rather than design an entirely new tank, the Germans should have focused on improving the Pzkw IV by giving it a more powerful and fuel efficient diesel engine, wider tracks and enhanced protection, but only from the most common allied AT weapons. As far as armament is concerned, the vehicle's KwK 40 L/48 gun was more than adequate. Germany's other excellent, more potent (but heavier and larger) AT guns could have been left to turretless tank destroyer/assault gun designs that could also be more heavily armored than conventional tanks due to the weight saved by mounting the gun in a casement.

Such changes would have allowed the Germans to produce far more AFV than they did historically – as an example, for each Tiger produced, the Germans could have manufactured seven Stug IIIs. Such numbers could have provided many more German infantry divisions with their own battalion of this very useful vehicle, and it could have even been distributed to Germany's allies in greater quantities and earlier than they were historically, which would have greatly improved these formations' staying power in the field.

If a new vehicle was to be made, it should have been an all-terrain truck that was capable of keeping the panzer spearheads supplied. The need for such a vehicle was obvious after the failure of Barbarossa – the German blindness to this lesson was one of the reasons for the failure of their summer offensive in 1942 as the panzers continually ran out of fuel and spares, causing tanks to be left behind and depleting their divisions of the mass needed to carry through their attacks and exploitation successfully.

The failure to appreciate the nature of the problems that beset their organization led to the downfall of the Heer in 1944 – 45. These final years were the most destructive of the war, so it is for the best that the German armed forces were incapable of prolonging the inevitable.

Normandy was the Wehrmacht's last chance to score a meaningful operational victory. If the allied bridgehead could have been rapidly driven back into the sea, significant reserves could have been sent east to stem the ruinous offensives launched by the Red Army in mid-late 1944. It is doubtful if the war would have had a positive outcome for Germany, but it would have made the allied victory a much more costly affair.

That the war was to continue for another year was due as much to allied logistical limitations as to German resistance. The Volksgrenadier Divisions, Volkssturm, Panzer Brigades and other expedient units could only briefly delay the inevitable – the Soviet Vistula-Oder Offensive annihilated scores of these stop-gap units in a matter of weeks.

The counter-attack around Arracourt, the Ardennes counter-offensive, Nordwind, and Operation Konrad and Spring Awakening were all doomed to operational insignificance due to the German's inability to logistically sustain them, the steady loss of mass in the Panzer formations due to increasing mechanical breakdowns and increasingly effective allied counter-measures.

The July Plotters who bravely sought the destruction of the Nazi regime were correct in their assessment of the conflict's progress – the war was effectively over and the only thing remaining to be determined was how much of Europe and her population were to be spared before it was over. The best that can be said of the Tiger and Panther tanks is that they helped hasten this conclusion.

Bibliography:
From Defeat to Victory – the Eastern Front, Summer 1944 by C.J. Dick
Tank Warfare on the Eastern Front 1943 - 1945 by Robert Forczyk
Armored Champion – The Top Tanks of World War II by Steven Zaloga
The Infantry's Armor – The U.S. Army's Separate Tank Battalions in WWII by Harry Yeide
Tank Warfare on the Eastern Front 1941 – 1942 by Robert Forczyk
Armored Thunderbolt – the U.S. Army Sherman in World War II by Steven Zaloga
Objective Ponyri! The Defeat of XXXXI Panzerkorps at Ponyri Train Station by Martin Nevshemal
Demolishing the Myth: The Tank Battle at Prokhorovka, Kursk, July 1943: An Operational Narrative by Valeriy Zamulin
The Battle of Kursk by David Glantz
Sledgehammers: Strengths and Flaws of Tiger Tank Battalions in World War II by Christopher Wilbeck
Tigers in Normandy by Wolfgang Schneider
Armageddon in Stalingrad: September to November 1942 by David Glantz
The Battle for Leningrad by David Glantz
Caen 1944 by Ken Ford
D-Day 1944 – Sword Beach and the British Airborne Landings by Ken Ford
Operation Totalize 1944 – the Allied Drive South from Caen by Stephen Hart
Panther vs T-34: Ukraine 1943 by Robert Forczyk
Anzio 1944 by Steven Zaloga
Stalin's Favorite: The Combat History of the 2nd Guards Tank Army Vols. 1 and 2 by Igor Nebolsin
Atlas of the Eastern Front 1941 – 45 by Robert Kirchubel

www.ingramcontent.com/pod-product-compliance
Lightning Source LLC
Chambersburg PA
CBHW080902220526
45467CB00008B/2600

* 9 7 8 1 7 1 7 8 7 1 1 3 8 *